Amazing Mammals

WRITTEN BY
ALEXANDRA PARSONS

PHOTOGRAPHED BY
JERRY YOUNG

Stoddart

Editor Scott Steedman
Designers Ann Cannings and Margo Beamish-White
Senior art editor Jacquie Gulliver
Editorial director Sue Unstead
Art director Anne-Marie Bulat

Special photography by Jerry Young
Illustrations by Ruth Lindsay and Julie Anderson
Animals supplied by Trevor Smith's Animal World
Editorial consultants
The staff of the Natural History Museum, London

Published in Great Britain by Dorling Kindersley Limited
9 Henrietta Street, London, England WC2E 8PS

First published in Canada in 1990 by Stoddart Publishing Co. Limited
34 Lesmill Road, Toronto, Canada M3B 2T6

Canadian Cataloguing in Publication Data
Parsons, Alexandra
Amazing mammals

(Amazing worlds series)
ISBN 0-7737-2398-6

1. Mammals - Juvenile literature. I. Title.
II. Series: Parsons, Alexandra. Amazing
worlds series.

QL706.2.P37 1990   j599   C90-093044-6

Color reproduction by Colourscan, Singapore
Typeset by Windsorgraphics, Ringwood, Hampshire
Printed in Italy by A. Mondadori Editore, Verona

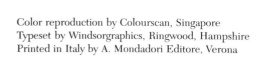

Contents

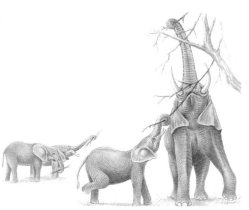

What is a mammal?

You are a mammal. Mammals are animals that are warm to the touch because they have warm blood. Most mammals give birth to live babies – not eggs – and the mother produces milk for her babies to drink.

All mammals – even whales – have some hair.

As big as a house...
The biggest mammal of all is the blue whale, which is about the size of a house – a big house – and as heavy as 2,000 people!

...smaller than a mouse
The smallest mammal is the rare bumble-bee bat. With its wings outstretched it's only half the size of the blue whale's eye!

What a team!
There are thousands of different kinds of mammals. Some of them live in the water; most of them live on land.

lion

sea lion

hare

zebra

armadillo

You are a primate (PRY-mate), one of a group of mammals that includes monkeys and gorillas. Some primates are very clever and learn quickly – thanks to their big, well-developed brains.

Slow starter

Human babies need their parents for much longer than any other mammal except the elephant. Humans can't walk or talk properly until they are one or two. Then they have to go to school to learn to read and write. By the time they've learned how to look after themselves, they are about 16 years old.

Shake those bones!

All mammals have a skeleton – a frame of bones inside. The skeleton is held together by the backbone. Animals with backbones are called vertebrates (VER-tuh-brates) because the little bones that make up the backbone are called vertebrae (VER-tuh-bray).

kangaroo

giraffe

wolf

peccary

beaver

The elephant

Elephants are big – very big. They are the largest land animals of all. A full-grown elephant is taller than a bus – and weighs as much as a van full of people!

six-year-old

newborn calf

Indian elephant

Elephant tusks are actually long teeth.

Big ears
Elephants live in India and Africa. Indian elephants flap their big ears to keep cool in the hot jungles. But Africa can get even hotter, and an African elephant's ears are even bigger.

The most amazing part of an elephant is its trunk – which is really a very long, all-in-one nose and top lip.

Nose pipe
Elephants use their trunks for all kinds of things. They smell with it. They grasp food and touch each other with it. They squirt water into their mouths and over their backs with it. They even use their trunks to trumpet loudly.

triarch

teenage
male

Women only
Female elephants and their children stay close together in family herds. Everywhere they go, they are usually led by the biggest female, called the matriarch. Adult males spend most of their lives alone.

The elephant's wisdom
Elephants were once thought to be very wise. In ancient Siam, people believed that if you whispered a problem in the ear of an elephant, the elephant would give a sign to tell you how to solve it.

Tree trunks
Elephants use their trunks to tear off leaves and branches and rip up clumps of grass, which they eat. Sometimes they rip up whole trees, roots and all!

The big toenail of an adult elephant is larger than your entire hand.

The camel

What a strange-looking beast! But nature has a reason for everything. The camel lives in deserts where there is little to eat or drink, and it has the ability to go for a long time without food and water.

Where do I sit?
When riding a two-hump camel, you sit between the humps. On a one-hump camel you perch in a saddle on top of the hump.

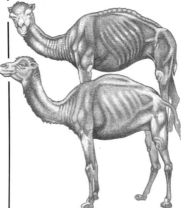

Thirst – no problem!
A camel rarely gets thirsty. Its body uses so little water that it can go for 10 months without a drink! The camel shrivels to a bag of bones, but one big drink and it is back to normal.

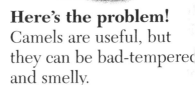

Here's the problem!
Camels are useful, but they can be bad-tempered and smelly.

Work – no problem!
Without camels, life for desert people would be hard. Camels carry heavy loads and provide milk to drink and thick wool to wear.

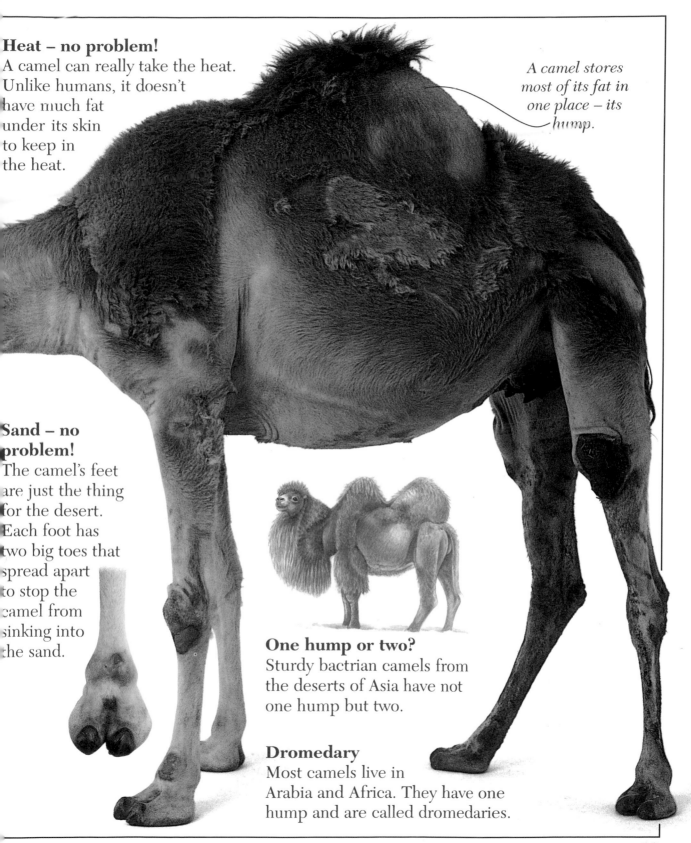

Heat – no problem!
A camel can really take the heat. Unlike humans, it doesn't have much fat under its skin to keep in the heat.

A camel stores most of its fat in one place – its hump.

Sand – no problem!
The camel's feet are just the thing for the desert. Each foot has two big toes that spread apart to stop the camel from sinking into the sand.

One hump or two?
Sturdy bactrian camels from the deserts of Asia have not one hump but two.

Dromedary
Most camels live in Arabia and Africa. They have one hump and are called dromedaries.

13

The fox

Foxes are members of the dog family. They sleep all day in underground burrows and come out at night to search the woods and fields for food.

Secret stores

Foxes don't hunt for food only when they are hungry. Sometimes they hide food in a secret place in case they get hungry later.

The mouse leap

Mice are tricky to catch. The clever fox sneaks up softly. Then it jumps up high and dives down right on top of its prey.

Foxed!

Foxes are famous for their cunning. An old story tells how a fox tricks a crow into dropping a piece of cheese. The fox flatters the crow and asks it to sing. When the vain bird opens its beak, out pops the cheese.

A fox's ears are so sensitive it can hear a worm wriggling on the other side of a field.

Mmm, tasty!
Foxes eat just about anything, from worms to chickens. Some even sneak into town and raid trash cans.

Arctic fox
Not all foxes are red. One kind of fox lives in the Arctic, where its white winter coat is a perfect disguise in the snow.

Foxy sounds
Foxes yap, howl, bark, and whimper, just like dogs do.

Family life
A mother fox gives birth to around five little cubs at one time. She looks after them while the father fox goes out to search for food. Both the mother and the father teach their cubs to hunt and fight by playing games with them.

A flying fox?

With its furry snout and big eyes, this strange creature looks like a little fox. But it is not. It is a fruit-eating bat or "flying fox," and it can fly as well as a bird.

Fruit detector
Some Mexican fruit bats love ripe bananas so much that Mexican farmers have to harvest their crop before it ripens. But the bat's sharp nose can even smell out ripe bananas inside a farmer's hut!

Bat club
Most bats are creatures of the night. During the day, they hang upside down in trees and caves, snoozing, chattering, and cleaning their wings. From a distance they look like closed umbrellas.

Juice extractor
A fruit bat has a mouth like an orange squeezer. The bat uses its tongue to squash fruit against special horny ridges in the roof of its mouth. When all the juice is squeezed out, it spits out the skin and pits.

Flying foxes grab fruit without landing.

Going batty
The world is full of bats – nearly 1,000 different kinds altogether. This Egyptian fruit bat is just one of more than 170 different kinds of flying fox.

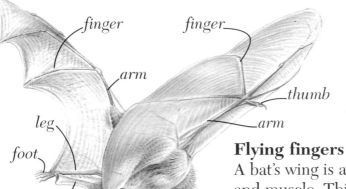

finger *finger*
arm
thumb
arm
leg
foot
tail

Flying fingers
A bat's wing is a thin, see-through layer of skin and muscle. This layer is stretched tight between the bat's bony legs, arms, tail, and long fingers.

Earsight
Most bats eat insects, not fruit. These bats can chase their prey in the dark by using their ears. They make very high squeaks and listen to the way the squeaks bounce back from nearby objects.

Vampires
The evil Count Dracula – who sucks people's blood – has given vampire bats a bad name. Vampire bats do drink blood, but usually from cows and pigs, not humans. They do live in old castles, though.

17

The sloth

Sloths spend most of their lives hanging upside down in trees in the rain forests of South America. They move very slowly and wake up only at night.

What is a sloth?
The word *sloth* means someone who is very, very lazy and slow – just like the sloth, in fact.

Too lazy to wash
Sloths don't clean their fur. After a while it grows a green scum, and moths and beetles come to live in it. This is a good disguise – it makes the sloth look like a bunch of leaves.

What a drag!
Sloths' legs aren't made for walking. If they ever find themselves on the ground (which doesn't happen very often), they have to drag themselves along with their claws.

Going for a swim

They're not good at washing, they're not good at walking, but sloths are surprisingly good at swimming in the jungle rivers.

The sloth's claws are like huge hooks, good for hanging from branches.

The sloth's fur grows backward so that water can drip off it.

Upside-down babies

Mother sloths even give birth to their babies hanging upside down. The baby crawls into the fur on its mother's tummy and searches for a nipple. It stays there for up to nine months, drinking milk, until it is ready to start hanging upside down all by itself.

The otter

Otters are more at home in the water than on land. Most of them live in rivers and streams, and one kind spends its whole life in the sea. All otters are good swimmers with long, slender bodies and strong tails.

Otter nonsense

Otters are very playful animals. When they are not hunting, they spend hours play-fighting, diving, and sliding down muddy banks into the river.

Funny feet

The back feet of an otter are webbed and look just like flippers.

Sweet dreams

Sea otters even sleep in the water, lying on their backs. The otter wraps itself in seaweed so it won't be swept away while it is snoozing.

Warm fur

This is a river otter from Asia. Like all otters it has soft, silky fur. The fur traps bubbles of air to keep the otter warm.

On your mark, get set...
Otters are fast and graceful swimmers. A sea otter can swim at 10 miles per hour – that's faster than you can run.

Setting the table
Sea otters eat floating on their backs. They use their chests as tables and often balance a flat stone like a plate on their chests. They smash clams and crabs on the stone to open them.

Unlike most water animals, the otter doesn't have a layer of fat under its skin for warmth.

Big appetites
Otters eat a lot of food, mainly fish and shell-fish. They use their long whiskers to find their food on the bottom of the river or sea. When an otter catches a huge fish, it may hold the fish to its chest with its front feet, then take it to the bank to munch at its leisure.

The squirrel monkey

This pretty little monkey lives in big groups in the jungles of South America. As many as 500 monkeys may travel together through the treetops, leaping from branch to branch like acrobats in the circus.

All dressed up
Squirrel monkeys have always been popular as pets. It was once fashionable to have hats and coats made for pet monkeys. Then people would walk down the street with their pet sitting on one shoulder.

The monkey uses its long tail for balance as it walks along branches.

All fingers and thumbs
The squirrel monkey uses its hands like hooks to grab on to branches. It has long, strong fingers. But its thumbs are so far up its wrists it can't use them for grasping the way we can.

For its size, the squirrel monkey has the biggest brain of any animal – including humans!

Drinking and dozing

Baby monkeys cling to their mother's fur and ride around on her back. They spend the first weeks of their lives drinking milk and sleeping, just like human babies.

Monkey menu

Squirrel monkeys live on leaves, fruit, and birds' eggs, spiced with insects and spiders and the occasional bunch of flowers.

Fur wash

Squirrel monkeys are very fussy when it comes to personal hygiene. They are always licking and brushing their fur. The monkey cleans its legs, stomach, and tail with its fingers, and uses its toes to comb its back, head, and behind its ears.

The porcupine

This spiky-looking creature is really a kind of overgrown guinea pig. It has sharp spines, or quills, and very long whiskers.

Prickly babies

Baby porcupines are born with bristles that harden into sharp quills within a few hours.

Gripping tails

Unlike their African cousins, these American porcupines spend a lot of time in the treetops. They grip the branches using their long tails and strong claws.

Good manners

Porcupines are neat eaters. They hold their food against the ground with their front paws and chomp into it with their big, sharp teeth.

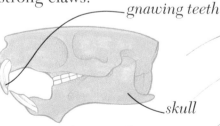

gnawing teeth

skull

Long in the tooth

The porcupine is a rodent. This means it has long, sharp front teeth for gnawing. As quickly as the teeth wear down they grow again. In fact, they never stop growing!

This African porcupine uses its whiskers to find its way in dark burrows.

Ouch!
Porcupine quills come loose and will stay stuck in the skin of an enemy. If germs get into the wound, even an animal as big as a lion may die.

Shake, rattle, and stab
When in danger, a porcupine will turn its back on its attacker, rattle its quills, grunt, and stamp its feet. If all this bluffing doesn't work, it reverses into its enemy, quills first.

A porcupine will rattle the hollow quills on its tail to scare off an attacker.

Punk hairdo
The porcupine's quills are specially adapted hairs that stand up without hairspray!

GEL

25

The tiger

Tigers are the largest of the big cats. This little cub is only ten weeks old, but when it grows up it will be a powerful hunter. At one time tigers nearly died out, as thousands were killed for their beautiful skins. Today all tigers are protected by law.

In for the kill
A tiger begins searching for its prey by sniffing the air. Once it can see its victim, the tiger creeps up stealthily until it is close enough to charge. It attacks from the side or behind, slashing with its paws and then killing with a quick bite to the throat.

Do tigers eat people?
Generally, no. But if a tiger is old or sick and too slow to catch its usual prey, it may start attacking people.

Blind and toothless
Like kittens, tiger cubs are born blind and without teeth. Their stripes are lightly colored and will get darker and darker as they grow.

Tiger tots

While the mother tiger looks after her cubs on her own, the father tiger prowls his range, warning other tigers to keep away.

In black and white

White tigers are rarely found in the wild but are very common in zoos.

Water sports

Tigers are strong swimmers. If a tiger cannot find enough large animals to eat, it will head for a river and snatch fish and frogs from the water.

Like all cats, a tiger can pull in its claws – until it needs them for catching its prey.

Hide-and-seek

The stripes on the tiger's coat blend in with the sunshine and shadows of the forest. It can sneak up on its prey quietly without being seen.

Bringing up baby

Most baby mammals start to grow inside their mothers and stay in there until they are ready to be born. Then their parents care for them until they are old enough to look after themselves.

A bear that's not a bear

The koala is a very cuddly little animal. It looks like a bear, but isn't a bear at all. It's really a marsupial (mar-SOUP-ee-ul), a kind of mammal that carries its babies in a pouch on its belly. Like many other marsupials, the koala is found only in the wild in Australia.

Hanging out with Mom

When a koala baby is born, it is hardly developed at all. It is pink and wrinkly and as small as a pea. The baby crawls into its mother's cozy pouch, latches on to a nipple, and doesn't let go for five months. It grows quietly, drinking milk and sleeping. When it is big enough, the baby koala starts to hop in and out of the pouch.

mother giraffe with calf

A mammal that lays eggs

The platypus looks like a small otter with a duck's bill for a nose. It is actually a bizarre kind of mammal that lays eggs. Platypuses are very primitive. They may have been around 70 million years ago, when dinosaurs still walked on Earth.

Mother's milk

Mammals like the platypus and the koala are unusual. Most mammals give birth to live babies which don't need a pouch. The first thing they do, other than cry, is look for a nipple and have a good drink of milk.

A pregnant elephant has to wait 22 months before her baby is born. That's almost two years – longer than any other mammal.

	Usual number of babies	Time baby is inside mother	Age baby stops drinking mother's milk	
Human	1	9 months	6-9 months	
African elephant	1	22 months	3-4 years	
Dromedary camel	1	14 months	9-18 months	
Red fox	4-6	7 weeks	8-10 weeks	
Egyptian fruit bat	1	4 months	3-5 months	
Linne's two-toed sloth	1	6 months	3-4 weeks	
Asian short-clawed otter	3	2 months	3-4 months	
Black-capped squirrel monkey	1	5 months	5-6 months	
African porcupine	1-3	4 months	4-6 weeks	
Indian tiger	3-4	15 weeks	3-5 months	
Koala	1-2	7 weeks	6-12 months	

Index

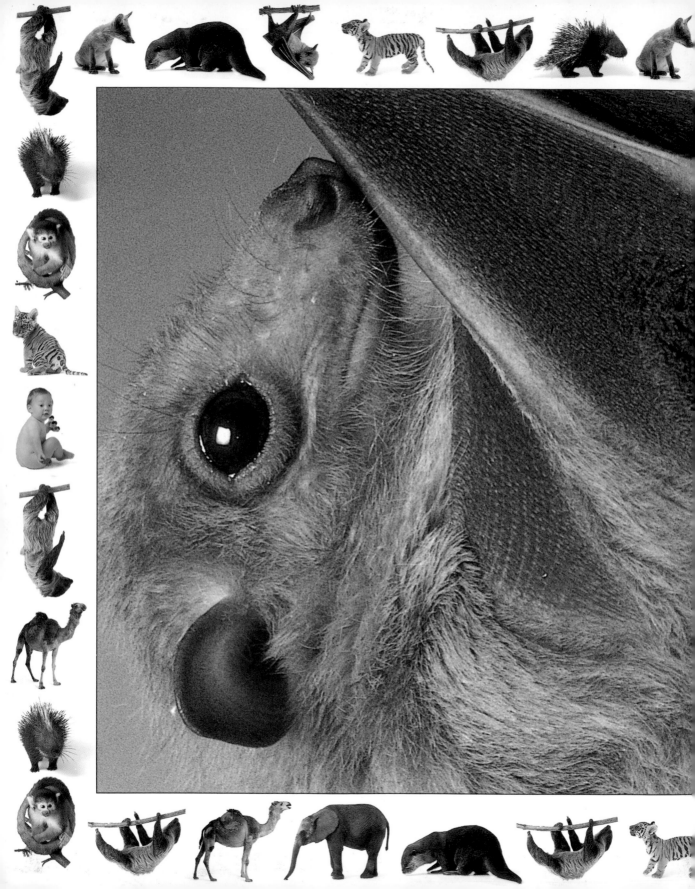